How The Old Fella Done It

HOW THE OLD FELLA DONE IT

Building the Shelburne Dory with Milford Buchanan

Dory Heritage Project

Authors and Researchers:

Brad Dimock • Douglas Brooks • Graham McKay • Cricket Rust

Text and images © 2025 by Dory Heritage Project

Text and photographs, unless otherwise credited, are by Douglas Brooks, Brad Dimock, Graham McKay, and Cricket Rust, with additional photographs by Diane Hirabayashi Carter.

Protected by copyright under the terms of the International Copyright Union; all rights reserved. Except for fair use in book reviews, no part of this book may be reproduced for any reason by any means, including any method of photographic reproduction, without permission of the author.

First Edition

ISBN
Hardcover: 978-1-892327-24-6
Softcover: 978-1-892327-23-9

LCCN 2024925574

Dory Heritage Project
Douglas Brooks, Brad Dimock, Graham McKay, Cricket Rust

Fiscal agent
Lowell's Boat Shop
Amesbury Massachussets

To Milford Buchanan, who has dedicated a quarter century and more than a third of his life to preserving the Shelburne Dory tradition.

Contents

Acknowledgments	xi
Preface	xiii
Milford Buchanan and the Shelburne Dory	1
Building the Shelburne Dory	11
Conclusion	41
Bibliography	43
Dory Shops	45

Acknowledgments

We owe tremendous thanks to Milford Buchanan for keeping the tradition alive, showing us the way, and putting up with our endless questions.

Many thanks to Milford's right-hand man Mick Fearn, for his decade-and-a-half of volunteer work helping Milford produce Shelburne dories.

Many thanks to Shelburne's Museums by the Sea, and to Suzanne Mahaney, Kim Robertson Walker, and Mike Hartigan for their support of this project. To Brian Ogilvy for his work researching and creating the interpretive material in the museum section of the shop. The information is well organized and invaluable to understanding the tale of Shelburne's boats and boatbuilders.

Thanks to Lowell's Boat Shop for providing the umbrella infrastructure for the fundraising and bookkeeping.

Thanks to the Traditional Small Craft Association for their generous grant in support of this endeavor.

Thanks to Diane Hirabayashi Carter, who took many of the photographs for this book.

Thanks to George Kirby Jr. and Sons Paint Company for clothing our team.

And a huge thanks to the many generous donors of our fundraiser in early 2023. Without you all this would not have happened.

Preface

The dory as we know it, in shape and name, evolved in and around Lowell's Boat Shop in Amesbury, Massachusetts in the mid-1800s. The basic design soon spread from Cape Cod to Newfoundland. Due to its perfect utility for the cod fishing industry on the Grand Banks, it became ubiquitous along the eastern seaboard. Each boatbuilding locale adapted the designs and methods, establishing their own unique tradition.

The 1900s brought new materials to boatbuilding—steel, aluminum and fiberglass rapidly crowded wooden boats off the market, their requisite methods and designs rapidly headed toward extinction. Overfishing and more modern techniques crushed the demand for dories.

If not for the efforts of a few dedicated individuals, most of what we now know of wooden boats could have been lost. Howard Chapelle wrote several books documenting designs and techniques. Maine's John Gardner took it upon himself to document countless small wooden boats along the New England coast, and collected many seminal specimens for the Mystic Seaport Museum collection.

Gardner's 1978 treatise, "The Dory Book," documented two dozen variations of the basic dory: some unique oddballs and some production models, the classic Banks Dory being one of them. Nearly every boat he described was already a relic. In the forty-five years since, the dory has nearly vanished.

In 1984 Newfoundland dory historian Otto Kelland described seven varieties of the Banks Dory that were in production during the heyday of cod fishing. To the layman they may have been indistinguishable, but to the fishermen the differences were often profound. Practitioners of these different varieties are now narrowed down to few or none, and the traditions may soon fade entirely without documentation now.

In 2022 a group of four of us gathered our collective interests and abilities to begin documenting the remaining dory traditions. The Dory Heritage Project was born under the umbrella of Lowell's Boat Shop. We are crowd funded with a few major donors and grants.

Graham McKay, Executive Director of Lowell's Boat Shop, is the leading authority on the heritage of the Lowell's dory tradition, and teaches the techniques. Boatbuilder Douglas Brooks has spent much of the last thirty years apprenticing and documenting aging Japanese boatbuilders, each the last of their lineage. Brad Dimock and Cricket Rust operate Fretwater Boatworks, building the Oregon-style river dory, and teach the techniques around the country. Together the four of us have a shared passion for the heritage and traditions of all things dory and have formed the Dory Heritage Project. Documenting Milford Buchanan and the Shelburne Dory is the beginning.

Milford Buchanan and the Shelburne Dory

How the Old Fella Done It

Brad Dimock

August, 2014

Milford Buchanan and I had bent the floor of the dory into the building horse—an ancient cradle on the dory shop floor that has birthed more than fifty thousand dories in its life. We had wedged the dory's floor down tight to the cradle with the same two sticks and jacks that had bent countless floors before. All five ribs were already attached, as were the stem and transom. The skeletal boat—at this stage called a skillet—resembled the bones of a primordial sea creature. The floor now had 3½" of rocker. "Up in Lunenburg they give it 5½", says Milford, shaking his head.

It was time to plumb the stem and secure it. I grabbed a level and held it alongside the stem. Milford looked at me, paused, and said, "I guess you could do it that way." He took the level from me, set it on the workbench, walked to the end of the shop, and shut one of the bay doors. He returned to the bow of the boat. "See that stick on the edge of the door? Sidney Mahaney put that there. That's plumb." Straddling the dory and squinting at the stem and the stick through one eye, Milford directed me to nudge the stem plumb and secure it. That's how it's always been done here at the Shelburne Dory Shop. Innovation is not terribly welcome here—the object is to preserve the tradition as best as possible, to do things the way he was trained, or as Milford frequently says, "how the old fella done it."

As a builder of river dories, I became intrigued by the enchanting lines of dories of all kinds. I was on my way back to Maine in 2014, having just visited the Dory Shop 100 miles east in Lunenburg when I stumbled across Shelburne's Museums by the Sea with its own Dory Shop, and Master Dory Builder Milford Buchanan. It was an unexpected treat. Although I only spent one afternoon working with Milford, the antiquity of the shop and Milford's dedication to tradition fascinated me—and later came to haunt me. A few years later I took a Japanese boatbuilding course from Douglas Brooks, building a Shimano River Boat. His tales of apprenticing under, and documenting, elderly Japanese boatbuilders, each the last of their line, enthralled me, and his spectacular book documenting these men, their tradi-

Milford using age-old bottle jacks and board to bend the skillet into the curve of the horse. August 2014.

The Dory Shop in 2014. Second dory from the left is a Gartside Picnic Dory.

The spring of 2023 saw the worst fires in Nova Scotia's history. Here the Jordan Bay fire is exploding near Shelburne. The Dory Shop is the small, dark building below and to the left of the steeple.

tions, and these boats moved me. I kept thinking of Milford, also the last of his lineage and the sole preserver of the Shelburne Dory tradition.

In 2019 I returned to Shelburne to work with Milford for a week, and wrote a short article for WoodenBoat Magazine. But it was perfunctory and incomplete.

I later met Graham McKay, master dory builder and Executive Director of Lowell's Boat Shop—birthplace of the Banks Dory tradition. He was intrigued with my tales of Milford and the Shelburne Dory. He agreed we needed to do a full documentation and, together with Douglas Brooks and my boatbuilding protegé Cricket Rust—also a dory devotee—we resolved to return to Shelburne and do it right.

We spent the better part of a year organizing ourselves, coordinating with Milford and Museums by the Sea, and fundraising. In late May, 2023 we caravanned to Shelburne, deviating around the major wildfires besetting Nova Scotia—the worst fires in Nova Scotia's history. The town of Shelburne and surrounding areas were under imminent evacuation readiness for the first week of our project.

A second shocker came when we met in the boatshop to begin work Monday morning: Milford announced this would be his final year, and the boat we were to build might be the second to last authentic Shelburne Dory ever built. We had nearly missed this opportunity.

History of the Shipbuilders and Region

Shelburne, Nova Scotia exploded into existence in 1783 with the arrival of some ten thousand British Loyalists from the newly independent United States of America. They had sided with the losing team and, no longer welcome in the USA, had headed for friendlier territory. Shelburne commands one of the world's great harbors, and it seemed

Shelburne circa 1897. The Dory Shop is near the left side of the photo, with a small light dock and a light roof. The sidewheeler City of St. John regularly plied the coast between Halifax and St. John. *Shelburne County Museum E-10b*

like an ideal place for a new city. A year later Shelburne was the fourth largest city in North America with an estimated population of seventeen thousand. Among those immigrants was a German militiaman, Conrad Bohannan, who settled at Jordan Falls, near the mouth of the bay, closer to the fishing grounds.

Reality reared its head within a few years when it became apparent that the rocky, spare soil of the southwestern Nova Scotia coast could not support enough crops to make a living. Most of the Loyalists moved on and the town imploded to under a thousand. Bohannan stayed. The family name anglicized to Buchanan and his progeny fell into the one line of work that was dependable—boatbuilding.

By the mid 1800s a revolution in fishing techniques was underway on the Grand Banks. Newer deep draft schooners were better able to withstand the notorious storms of the Banks, and cod fishing was highly profitable. It was readily apparent that hand lining from the schooner rails was not the most efficient way to fill your hull with fish. Trawl fishing from a fleet of boats was the answer. Enter the Banks Dory.

The Banks Dory is the archetypical dory—the classic fishing dory of Winslow Homer paintings. It evolved in the early to mid 1800s in and around Gloucester, Essex, and Salisbury, Massachusetts. Lowell's Boat Shop in Amesbury, the oldest working boatshop in the United States, is the consensual locus of the dory-as-we-know-it's birth. Simeon Lowell founded the shop in 1793, and began creating flat-bottomed boats with rounded sides. Simeon's grandson Hiram Lowell is often credited with the classic Banks Dory's "invention," although others may well have been migrating to fill this new niche.

What was needed was a cheap, simple, seaworthy, quick-to-build boat that could be stacked or nested on

Hiram Lowell of Lowell's Boat Shop is generally credited with formalizing the design for the Banks Dory, of which the Shelburne Dory was a variation. The Lowell family ran the shop for nearly two hundred years. It now operates as a nonprofit working museum. *Lowell's Boat Shop*

the decks of a schooner, and easily launched and retrieved. Form followed function. Like many other boats around the world—the French béte, the British Bridgewater Flatner, the Eskimo Umiak, the Oregon drift boat, and the New England bateau—this same classic shape emerged. A flat, slightly rockered bottom with flat, flared sides, sometimes with a slight transom. It fit the bill for Grand Banks fishing. It was simple, crude even. Yet incredibly elegant in its simplicity. Iconic. The Grand Banks Dory took off and became what is most likely the most-built rowboat in history.

According to Shelburne boat historian Lewis Robertson, local boatbuilder Isaac Crowell made an undated pilgrimage to "Essex and Salisbury Point to work with the master

Isaac Crowell is credited with bringing the dory design to Shelburne and inventing the "dory clip" which greatly simplified the building of dory frames. *Portrait by Thomas Doane, Shelburne County Museum 1967.54.01*

Walter Etherington learned the dory trade from Isaac Crowell and, in turn, taught John Williams. *Shelburne County Museum P-Eth1*

craftsmen of these towns. He secured and brought home plans for the most highly developed Gloucester type dory…" Salisbury Point being the home of Lowell's Boat Shop, it is likely he learned from one of the Lowells.

The Banks Dory design soon took root in boat shops from Massachusetts to Newfoundland and dories were produced by the tens of thousands. Like Darwin's finches, each dory building community evolved its own distinctive style. By the turn of the century the Banks Dory had, according to Newfoundland dory historian Otto Kelland, seven varieties, the Shelburne Dory being but one of them. Kelland, born and raised on the Newfoundland waterfront, in and around dories, recorded a gold mine of dory traditions.

Shelburne, with its extraordinary harbor, was a logical stop on the way from Massachusetts to the Grand Banks, and by the mid 1800s the town boasted ten shipyards and seven dory shops. Schooners heading for the Banks could put in here, pick up enough dories for the season, and head out. At eighteen dollars apiece, dories were considered a throw-away boat, expected to last a season or two. If a gunwale was crushed against the hull of a schooner when boarding in heavy seas, it might simply be set adrift, to be replaced at the next port call.

In 1880 John C. Williams—who learned from the Etherington family, who learned from Crowell— opened the largest production dory shop in Shelburne, employing seven men. They produced a dory a day, paint still wet as it slid out the door. Sometime around then Isaac Crowell introduced a metal clip system that eliminated the need for grown tree knees for the dory ribs. The dory frames could now be sawn from regular planking and quickly clipped together at the corners. Although the innovation was offered to other boatbuilders along the coast, craftspeople are slow to change their ways. But the Williams Dory Shop adopted the timesaver and production doubled.

John Williams and, later his adopted son Charles Wyman, ran the shop for nearly seventy years.

John Williams Boat Shop back when dories were big business. *Shelburne County Museum H-JC01c*

Sidney Mahaney built dories for seventy-six years, working at the John Williams Boat Shop until it closed in 1971. He returned for another ten years when the shop reopened as a museum. *Shelburne County Museum*

About one hundred years ago Williams's shop modernized with electricity and a Crescent Universal Wood Worker—a goliath belt-driven iron machine featuring a 32" band saw, table saw, jointer, shaper, and a wood borer. Production jumped again. (When the Wood Worker broke down a few years ago, former Dory Shop owner Bill Cox came by to supervise repairs. He was older than the saw. Cox, who Milford calls "the old fella" came by weekly to check on Milford until he passed at 104.)

Shelburne's most notable dory builder, Sidney Mahaney, started working in the JC Williams dory shop in 1914 as a young teen, making forty-five cents a day. He soon rose to the esteemed position of dory builder, hauling in a full dollar a day. For seventy-six years, Mahaney built dories and claimed to have built or worked on ten thousand of them himself. The Williams Dory Shop produced more than fifty thousand. Williams operated the shop for nearly sixty years until his death in 1939. His adopted son Charles Wyman ran it for another twenty. In 1939 brothers Bill and George Cox acquired the shop. Sidney Mahaney was still there.

But the dory's heyday was doomed with the introduction of diesel ships and draggers, which hauled huge nets along the floor of the Banks, catching tremendous numbers of fish but destroying the ocean floor habitat. The Banks fishery collapsed, although dories continued in demand as lifeboats for a few more years. In 1971 Bill Cox closed down the shop—not so much due to a lack of demand but, "Our old dory builders simply wore out." He could not get a younger generation to train in under the old masters. He simply closed the doors.

When Prince Charles and Lady Diana officiated at the reopening of the shop as part of Shelburne's Museums by the

Harley Cox (center) ran a major boatshop in Shelburne for many years. His sons Bill and George bought the John Williams Boat Shop in 1959 and ran it until it closed in 1971. *Shelburne County Museum, undated newspaper clipping*

Sea in 1983, Sidney Mahaney was back as its prime exhibit, building dories until his death in 1993. Mahaney's son Curtis then stepped in, along with Bill Cox, the last shop owner. When Curtis retired in 1998, Milford Buchanan, direct descendant of Conrad Bohannan, shirt-tail relative of Sidney Mahaney, and fourth generation boatbuilder, became the Dory Shop's Master Dory Builder.

I was born here in Shelburne. Sandy Point. 1956. My great-grandfather Tom—he was a ship's carpenter more 'n likely. My grandfather George, he used to fish out on the Grand Banks. He got lost. I was told that it

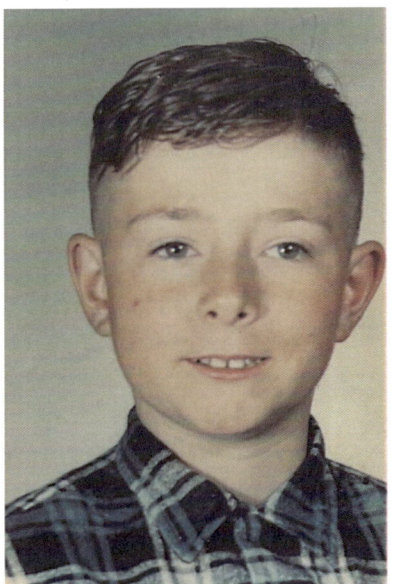

Young Milford, the ship model maker.

Milford at the furniture factory.

Bill Cox and Curtis Mahaney trained Milford in the art of the Shelburne Dory. Cox came by regularly to check on Milford until his death at 104 years old.

took him three days to get to shore. Both my grandfathers, Arthur Buchanan and George fished out on the Grand Banks. Arthur has a dory on his tombstone with a schooner in the background.

My father, we built boats and stuff. We lived out on this little point so we caught fish, lobsters. And dad was always building boats for my mother's brothers down in West Green Harbor, cause they were all fishermen. They'd want dories, skiffs, and stuff like that. And I used to help Dad and my grandfather dig the stumps out of the ground. This was a lot of work, you know? And dad always used hackmatack, we never used oak. In the early days we were building dories.

My grandfather George always used to model these great big two-masted schooners. And they'd be three or four foot long. I fell in love with that. It took me a long, long time to learn how to make those models. And so when I was a teenager instead of goin out and mowin grass and stuff, I went and made little ship models and sold 'em at the craft stores. And so today I have a business in which I make ship models and whirligigs. And I have ship models on a four-boat whirligig that spins in the wind. And the sails all tack. I've fallen in love with that ever since I was a kid. And I started that roughly when I was about eight years old. Mind you they weren't too good. They were rough lookin'.

After my father passed away, I retained his fishing license. And I couldn't remember how he done the dories and stuff. Now my cousin Sherman out on Enslows Point, I went out to him and I wanted to buy a boat and he laughed at me and said, "You guys have been building boats for generations, don't you think it's about time you learned how to build one for yourself?" And he said, "You've got that big cabinet shop in there, I'll come in in the next week and I'll get you started." I haven't stopped. I've been building them ever since.

And I'm also a cabinet maker, I worked in a furniture factory, as a foreman. And I did a little bit of carpenter work. But I liked building boats. I tried to be a Mountie three times, and I guess that wasn't to be. I was meant to be a boatbuilder. So I guess it's always been in my blood, the big ones or the little ones.

Bill Cox used to come out and stuff. He owned the dory shop. He took over from the previous owner who passed away. You know, one thing leads to something else. Cause he come out and said, "We're gonna need a new boatbuilder at the museum and I think you'd be our man." That would have been, let me see, 1998. He

The John Williams Boat Shop as it stands today, a part of Shelburne's Museums by the Sea.

didn't ask me, he told me. And he said, 'Now Milford, I want you to stay to this for at least twenty years. I don't want to train you and give you all this knowledge and you quit two years from now.' And back then you give your word with a handshake, and that's what we did. My father said, 'A Buchanan does not go back on his word.' And I didn't. I stuck right to it.

The Shelburne Dory

Dories are sized by their bottom length, because that's the part that actually gets built first and is the only part measured to length. The boat grows from there with ribs, stem, and transom. The side planking is wrapped on lapstrake, and fit to the framework piece by piece, but not measured per se. So a "15-foot" Banks Dory will be about 19–20 feet long up top.

The Shelburne Dory may look a lot like Lunenburg dories, or Gloucester dories, or Lowell Boat Shop dories—that archetypical dory shape, painted dory buff with green gunwales. But each locale made—and still makes— them a bit different. Each town still has one boat shop left that specializes in dories. When asked how his dories differ from those in Lunenburg, Milford says, "Our boats have a 3½" rocker. Lunenburg's have 5½". Ours has a bullnose cut on the bow and a ribbon (the rub rail or outer gunwale). "Ours are

The narrow stairway to the boatshop on the second floor.

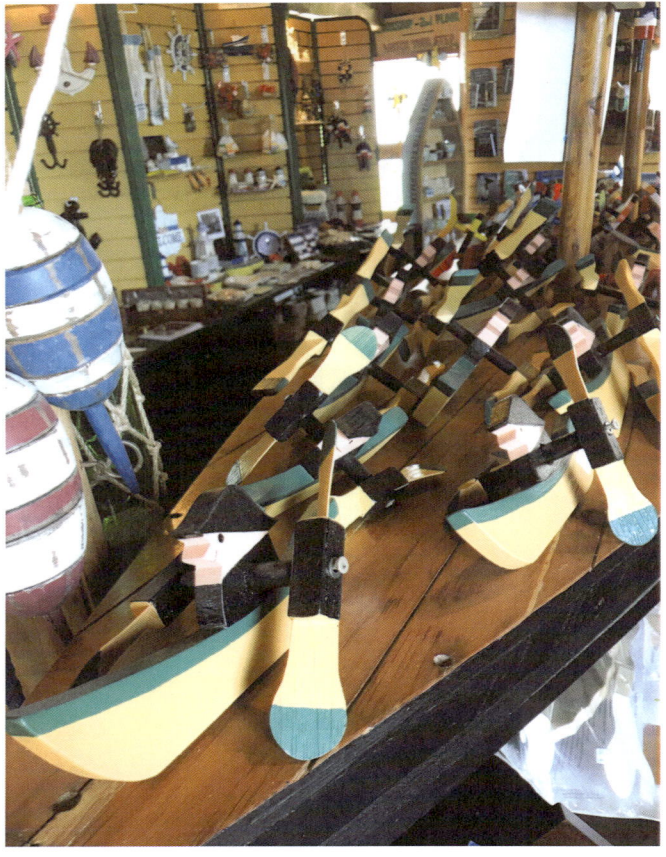

Milford-built whirligigs help support the museum, and Milford as well in the off season.

a fair amount lighter, due to somewhat thinner stock." Then he adds with a grin, "Theirs leak. Ours don't."

The dory workshop is precisely where John Williams founded it—on the second floor an old salt warehouse on Dock Street. The first floor is devoted to museum exhibits. Central posts supporting the second story prevented dory construction on the first floor. Besides, the second floor was not strong enough to support the huge supply of lumber that needed to be on site. So the boats were, and are, built upstairs and lowered out through large doors at the end of the shop. A narrow, worn staircase provides the only access to the shop. "It has to be that narrow," says Milford. "If any two people pass each other on the stairs, it's bad luck. Then you gotta get rid of the boat you're building or someone will die in it."

Production has slowed down a bit in the last hundred years, dropping from two or more per day to two or so per year. "We get an order for a boat most years," says Milford. The price has gone up a bit from the original eighteen dollars to anywhere from fourteen hundred dollars for a five-foot decorative coffee-table model, to nearly sixteen thousand for the twenty-foot Whaler, which measures 25'9" on top. The standard fifteen-foot Grand Banks dory runs $6750.

Nowadays Milford works slowly and carefully on each boat, pausing often to give tours of the shop or work on other projects. Milford is a master boat model maker as well,

Milford at work on a complex Picnic Dory with its reverse-curved wineglass transom.

and is the town's chief supplier of whirligig weather vanes. Sales of these help fund the shop between boat sales.

He is also a great entertainer, chatting up visitors, explaining the history of the boats and the shop, and describing what is going on with the current dory. He has a good supply of humor and a ready laugh. "Hello. I'm Milford, the Master Dory Builder here. How many dories would you like today?"

A few years ago renowned boat designer Paul Gartside, who used to live a few blocks from the dory shop, brought a new dory design for Milford to build—called the Picnic Dory, with complex curves and a graceful wineglass transom. He said it would have to be built upside down. "Not in my shop," responded Milford. "We've been building them right side up for four generations. That's not going to change." And he built it right-side up, with the curvy steambent planks and all. "The way things were handed down to me—that's the way I do 'em."

Not that everything is done exactly as it was in 1880. The dories are no longer expected to last just a season or two. Folks want to keep them for years, decades even. So the original method of nailing the floor together onto the ribs has given way to epoxy and biscuit joined floors that do not need to be recaulked each year. "People are used to fiberglass and aluminum boats that they don't have to caulk. They don't want to do that." The ribs, stem, and transom are now fastened to the floor with screws and a dab of 5200 instead of square boat nails, so they won't loosen and leak over time. A few power tools have crept into the system—cordless screwdrivers, a belt sander now and then. And of course the "new" hundred-year-old Crescent Universal Wood Worker's band saw, table saw, and jointer. But hand planes and eye are still the primary shapers, and the ancient patterns are still used for the floor and each piece of the framing.

Pine planking and oak for framing, transoms, and stems still comes from the Scott family, north of here, as it has for as long as anyone can remember. The Scotts are renowned champion log rollers, their clan dating back generations to when the log drives down spring-flushed rivers were the primary means of timber transport..

The galvanized square boat nails favored for dory building are no longer available. Fortunately the shop had a vast supply on hand when they closed in '71, and there are still plenty of nails for many dories to come. Paint, as for all traditional Banks dories, is Dory Buff—a mustardy yellow—with forest green gunwales. They say that the color combination stands out on the sea and in the fog.

The Crescent Universal Woodworker, still shaking the building after one hundred years.

A crate of boat nails remaining from the shop's final production years half a century ago.

Mick Fearn has volunteered as Milford's right-hand man for a decade and a half.

Mike Hartigan has been a regular visitor and helper in the Boatshop for many years. He is a wheelwright and a Director of the Shelburne Historical Society. His main focus these days is restoring the old wood mill a few hundred yards down Dock Street.

Fourteen years ago Mick Fearn and his wife were touring Nova Scotia and visited the museum. Mick had set a goal to retire at fifty and he did, handing his food equipment business in New England over to his employees. When Mick heard Milford's main helper was retiring due to health issues, Mick mentioned he'd love to lend a hand. They exchanged cards. Three weeks later Milford called, asking if Mick was serious. He was. He and his wife came back up for much of the season and ended up buying a summer home nearby. "At first it was a lark," says Mick. "But I eventually saw a need." Mick has volunteered as Milford's right-hand man for much of the summer season ever since. They don't cut each other much slack, and their banter could make good dialog for a road show.

When I met Milford Buchanan in 2014 he mentioned he had two apprentices, but they were both older than him and not likely to carry on the trade. "There's Mike Hartigan and Mick Fearn, that help me. They're not gonna be takin' over, no. I have one son, he's not even interested in woodwork, boatbuilding or nothin'. He has no desire. And I'm a fourth generation boatbuilder."

Who will carry on the legacy of the Shelburne Dory? When we asked again in 2023 the answer had not changed. "Nobody right at the moment," says Milford. "We keep talking about it, and I'm assuming it'll have to happen soon. 'Cause I'm retiring this fall."

The Shelburne dory tradition is easily traced. Isaac Crowell imported it from Massachusetts. Thence to Walter Etheredge and his son, to John Williams, to his adopted son Charles Wyman, and on to Sidney Mahaney, Bill Cox, Curtis Mahaney, and finally Milford Buchanan. I asked Milford where the bulk of his knowledge of the Shelburne tradition came from.

Well, it was mostly Bill Cox, and Curtis Mahaney. Now my father used to build 'em and stuff, but he didn't pass too much down onto me. 'Cause he didn't think I'd ever make a boatbuilder. You know? So I'm hopin' he's smilin' down on me right now.

Postscript: I dropped in to see Milford in late August, 2023, as his career was drawing to a close. Anne Poirier of Quebec had just arrived and is being trained to take Milford's spot in the future. Anne has been studying boatbuilding in Newfoundland. Milford has agreed to come in occasionally over the next year or two to help Anne master the basics of the Shelburne Dory the way the old fella done it. The tradition, for the moment, is hanging on.

Building the Shelburne Dory

Building the Shelburne Dory

Douglas Brooks

Entering the Dory Shop in Shelburne is a rare experience. Though now part of the larger institution of the Museums By The Sea[1], the shop defies conventions. While downstairs houses formal exhibits and displays, the second floor has almost nothing to suggest that you have entered any space other than a working boat shop. There are no barriers for visitors except in front of the large door where thousands of dories were launched from the shop. The floor is likely to be strewn with shavings and, as a centerpiece, there has regularly been a new dory under construction, its bright, fresh pine contrasting the floors, walls, ceiling, and benches, which have a deep patina from age.

According to Milford Buchanan, the shop's boatbuilder, the building was constructed by local shipwrights in 1865 as a salt shed. Like dories, salt was an essential commodity for the fishing vessels that worked from this harbor. In 1886 the building was converted to a boat shop. If one looks closely at the old timber frame of the building, you will see the builders used natural crooks instead of straight bracing where the posts meet the floor joists. These are called hanging knees in shipbuilding.

This use of shipbuilding techniques was neither quaint nor decorative. With seven shops on the waterfront building dories and small boats, along with several shipyards, no doubt there was plenty of material and obviously plenty of expertise. The shipwright builders may have gone against perceived conventions, but they knew how to join timbers. It is an example of what is called tacit knowledge, finding solutions based on experience and intuition.

Over the course of two weeks working under the direction of Milford Buchanan, we were to see many examples of this kind of knowledge. Since 1886 dories have been built in Milford's shop without any plans or drawings. Throughout the shop, hung on all the walls and stacked between the rafters are patterns for practically every part of the boats once built there. Except for a few copies Milford has made, all the patterns are dark with age, their edges black from the thousands of times they have been laid on planks, pencils tracing their perimeters. Some patterns are labeled but barely readable; most are so weathered and burnished no labels can be seen. Milford admitted he doesn't know what boats most of the patterns are for, and this would be difficult if not impossible to reverse engineer, since over the years the shop produced not only fishing dories but small skiffs, and lifeboats.

But Milford understands very well the patterns he has used since he started working in the shop in 1999. His teachers were Bill Cox and Curtis Mahaney. Bill had taken over the shop in the 1950s with Sidney Mahaney. Milford likes to point out the boatbuilding lineage of these three craftspeople can be traced back to the early 1900s. Throughout this entire history knowledge has been passed directly from the "old timers" to younger craftspeople. Now there is only Milford, and it's fair to say the techniques we attempt to document in this book run in an unbroken line back over one hundred years.

In our last conversation, Milford told us how his teachers demanded he commit to working at least twenty years in the shop as a precondition for being allowed in and taught how to build dories. Bill Cox and Curtis Mahaney recognized the value of what they knew but also insisted any new person

---

[1] https://www.shelburnemuseums.com

Patterns fill almost every inch of wall space throughout the shop, and nearly every part of all boats produced there was patterned.

respect the tradition. Milford made that promise to them and kept it. At the time we worked together, Milford was nearing retirement and no one had come forward to take over the dory shop from him, although in 2024 Milford has begun volunteering teaching a new builder. Recognizing the tenuousness of Shelburne's dory-making traditions and the unique opportunity to study with Milford, we have done our best to record his techniques as thoroughly as we can, but we take full responsibility for any errors and omissions in the text.

Materials

Before we joined Milford Buchanan in Shelburne he and his assistant Mick Fearn prepared our materials. The principal wood used in building the dory is Eastern white pine. It was sourced locally from a sawmill Milford has used for years and where the sawyer understands the type of material Milford needs. The planking stock was flat-sawn planking, also called flitch sawn, eighteen to twenty-two feet long. The planks retained their natural edges and therefore were tapered. The material was 16" at the widest. All framing in the dory, along with the stem, transom, strips, gunwales and

From left to right, Mick Fearn, Brad Dimock, Milford Buchanan, anc Cricket Rust look at the white pine prepared for the project. To the left is the material joined for the bottom.

cap rails is red oak. Milford acknowledged white oak is superior to red but the latter is much more readily available. This material is also sourced directly from a local sawmill. For the thwarts Milford purchased spruce from the lumberyard.

Milford mentioned many times the decline in lumber quality he has seen over the years. Principally this is in reference to how hard it has become to find clear, wide pine for planking. Most of the time Milford and Mick spent before our arrival was in drilling out the knots in the planking stock and filling with thickened epoxy. He uses a drill bit larger than the knot and drills it out to about ¼" in depth from both sides. His filler is a mix of epoxy and sawdust. Once the knots are filled he belt sands the filler flush to the plank surface. Milford said in the old days loose knots were either knocked out or drilled out and a pine plug was whittled to fit. The plugs were wrapped with a strand of cotton as caulking and driven into the hole tight, and a small finish nail was toe-nailed to hold them in place.

Milford said, "Up to the 1970s I used to be able to buy cotton caulking, galvanized boat nails, and everything I needed at the local hardware store." Increasingly over the years specialty items have become harder and harder to find and Milford said he feels very fortunate he can still source the wood he is using.

Bottom

The bottom of our dory is composed of five planks of straight-edged pine 8" wide. Milford likes to orient his planks with the heart side of the wood facing the outside. However, he said this is not a hard-and-fast rule. When choosing flitches later for the side planking he might ignore this. Milford and Mick assembled the bottom before our arrival as Milford was concerned our two-week schedule was not long enough to include this first step. Today Milford uses a biscuit joiner and glues the bottom together. First, he biscuits and glues the two outside pairs of planks. When the glue has set these two sections can be run through his wide planer and surfaced to 1⅛" thick. Next, the two pairs are biscuited and glued to the center plank and those seams are finished flat by hand. It is at this point knots are filled and sanded flush.

Traditionally the bottom planking was fastened with red oak floor timbers—which Milford calls strips[2]—and the bottom planks were nailed to them. In those days the planks were first clamped together and the bottom marked to shape before the strips were installed. Milford said the bottom planking was always square edged; there was never a caulking bevel. Instead, the builders used a caulking wheel to force a strand of cotton in the seams. He says older boats were built with four bottom planks, the center seam passing directly down the middle, acting as a physical centerline for the layout of stem, frames, and transom. He pointed out this created a leak problem where the seam passed under the stem and transom and sometime in the 1980s, he thinks, builders switched to five planks, which is how he has built all his dories.

Milford's method of gluing the bottom was the first of a few deviations from the methods he was taught by the old-timers in the shop. He was quick to point out he felt changes he made were always significant improvements. In the case of the bottom, gluing the planks ensures long-lasting, watertight seams. Also, the shop currently is not mass-producing dories for the fishing industry. In the old days dories were used daily and their bottoms were likely to stay swelled tight, plus the boats had relatively short lives. Today's customers are looking for recreational boats that might be used infrequently and kept out of the water for long periods of time. They want the boats to be convenient and long-lasting.

Milford says he likes to join the bottom boards leaving the seams tight on the ends and slightly open in the middle, with a gap of about ⅛". When the planks are glued and clamped this gap creates added pressure at the ends of the glue seams. Milford had previously worked in a furniture factory in Shelburne where among other things he built oak church pews. He says this was a technique he learned there, and it had been good training for his later boatbuilding because of the precision furniture-making demanded.

With the bottom assembled we began to lay out the shape. We always referred to this boat as a fifteen-foot dory, but among builders in New England and the Maritimes, the "length" of a dory was not its overall length but the length of the bottom. With a chalk line we struck a lengthwise centerline and then at the midpoint of fifteen feet we squared a line across it. On the squared line we marked points 18⅛" from the longitudinal centerline; this was the maximum beam of the bottom amidships. This squared line also marked the center of the middle pair of frames. Working toward bow and stern we marked off the centerline of the remaining frame locations at 29½" intervals. The names of the frames from bow to stern are: Forward, Forward Quarter, Middle, Aft Quarter, and Aft.

Milford had a pattern for the shape of the bottom. He said he thought it was one hundred years old. It was a narrow, curved piece of ¾" pine just under eight feet long.

---

[2] Other terms used for these timbers are floor or cleat.

Mick and Milford use a pattern to draw the shape of the bottom from the center to the ends. One edge represents the shape forward while the other the shape aft.

The curve of the two edges were slightly different: one was the shape of the forward edge of the bottom and the other the shape aft. With these curved lines drawn on the bottom we had a lozenge shape fifteen feet by just over three feet.

Although the edge of the bottom will have a strong bevel to receive the first side plank, Milford said in the old days it was always cut square-edged by hand and planed to a bevel only after the frames had been fastened. The bevels of the frames provide a guide for the bottom bevel. To save time Milford cuts the bottom bevel with his circular saw set at 27 degrees. The bevel still requires planing but the cut is about ¼" outside its final shape.

Frames

The construction of the frames in the dory represents one of the great innovations in dory building. The frames span both the sides and bottom, with their bottom sections lying alongside one another. Originally frames were made of natural grown crooks, milled from the intersection of the trunk and its roots. The curve of the grain provided the needed strength at the point where the frame turned from the bottom to the sides. Gathering frame material was difficult (finding crooks with the precise angle) and arduous (digging up tree stumps). Given the enormous production of dories being built in the 1800s to supply the fishing industry, boatbuilders were under pressure to find innovations that would speed production. Solving the issue of frame material would have been paramount.

Prior to its introduction, a shop with five to six workers, laboring ten hours, could produce a single dory a day. Once frames could be pieced together with clips production doubled to two a day. By 1900 clips were being mass produced by a factory in Ontario for the dory shops of Shelburne. Interestingly, in nearby Lunenburg the builders never adopted the dory clip. Even today The Dory Shop, which has been building boats in Lunenburg since 1917, builds its dories with natural grown crook frames.[3]

Each frame is made of two pieces of 1" thick red oak, a bottom frame and top frame.[4] The frame stock is 1¾" wide but increases to just over 3" where they join. A slight angle is cut on the edge so in cross-section each frame is a parallelogram. Once the frames are fastened to the bottom and it's bent to shape, all the frames should stand vertically. On the sides each frame needs to lay flat against the curve of the planking. Only the center frame is not beveled. One must be careful to get the angles right since the pieces only fit together one way (we mistakenly cut one frame backwards). We used the bandsaw and planed the bevels by hand. The bevels of the frames are as follows:

	BOTTOM	TOP
Forward Frame	6 degrees	18 degrees
Forward Quarter	3 degrees	9 degrees
Center Frame	0 degrees	0 degrees
Aft Quarter	3 degrees	10 degrees
Aft Frame	6 degrees	18 degrees

---

[3] The Dory Shop is located on the waterfront at 175 Bluenose Drive, Lunenburg, Nova Scotia. www.doryshop.com

[4] "Bottom" and "top" are Milford's terms, but they pre-date him as the oldest frame patterns in the shop are labeled this way. These are the terms used in this manuscript.

Milford's frame patterns showing the slight differences in angle of the overall frame. Milford used these patterns to set up the jig to fasten the frames together.

A bottom and top frame pair clamped tight against fences. This jig setup allows the builder to fit the joint and fasten the frames together, maintaining the proper angle.

Milford had a set of patterns for each frame for a fifteen-foot dory. He had carefully made new copies of these patterns from the old originals. In one corner of the shop is a jig with a pair of adjustable fences for making frames. The angles where the two pieces of the frame meet and are fastened with clips are first rough cut. The fences of the jig are checked with the frame pattern and tightened and then the frame material is clamped to the proper angle. Then a handsaw is used to make a cut through the joint, after which the two pieces are brought together again. Because the handsaw cuts a parallel-sided kerf, with each cut the fit at the bevel improves. When the joint is fit perfectly the two frame pieces are clamped in place and the clips are installed.

The frame angle for each pair is as follows:

	FRAME ANGLE
Forward Frame	123.5 degrees
Forward Quarter	124.5 degrees
Center Frame	124.5 degrees
Aft Quarter	123.5 degrees
Aft Frame	122.5 degrees

Milford saws through the joint between clamped frames.

The clips are prepared on another bench. Each frame joint is fastened using two pieces of galvanized sheet metal about $1/16$" thick. A pattern is used to mark for drilling and a set of holes is made in each pair. A $5/32$" bit is used to bore six holes through the clips while the frames are drilled slightly smaller at $9/64$". The first plate is fastened to one face of the joint with two nails, then the frame can be removed from the jig, turned over and the second plate driven on over the protruding nail heads on the opposite face of the joint. The remaining four holes are drilled and the nails installed. The

(Left) Dory clips await installation. They are 1/16" galvanized sheet metal. (Right) Nailing pattern for the clips, with nails driven from both faces. The tips of the nails will be clipped off then, with the assembly laid on a large anvil, the nail ends will be peened. The last step is hammering the overhanging edges of the clips, folding them over the joint.

points of the nails are clipped off leaving about ¼" of material. The frame is then laid on an anvil with the heads down and, using a heavy hammer, the tips of the nails are peened over. The final step is to hammer down the overhanging edges of the plates so they wrap around the frames.

It is crucial that the final assembly of the clips and frame be stiff and tight. Milford discovered one of our frames was loose at the joint and the only thing we could do, short of making an entirely new frame, was to continue peening the nails. It stiffened up to Milford's satisfaction.

Milford explained in the old days frames were fastened with "black iron" nails. Later 2¼" galvanized cut boat nails were used, always driven through the bottom from under the boat and into the frames. This was hard work, drilling holes and swinging a hammer working on the floor, and today because Milford's knees limit the amount of work he can do, he now uses 2" #8 stainless steel square-drive flat head wood screws, six spaced by eye under each frame. Milford further justified the use of modern screws saying, "We are trying to make a better boat."

The frames consist of pairs lying fore and aft on the lines we had squared off the centerline earlier. In every case the port side of the pair is forward, starboard side aft. "That's the way the old fellows always did it. I don't know why," Milford commented. In the course of our work, he mentioned many habits in the boat shop that were immutable. Tape measures,

Mick, Cricket, Milford, and Graham McKay set a pair of frames on the bottom.

for instance, were forbidden; boatbuilders could only use folding rulers. While reasons were often never given Milford continued to adhere to most, if not all, of these traditions.

The end of the bottom section of each frame was marked and trimmed by eye at about a 30-degree angle, leaving about a 1½" space for a waterway. Of the angle, Milford said, "As long as it's not square." Then the frame was used as a pattern to mark the length of its partner frame. The tops of each frame remained long and would not be trimmed until the boat was completely planked.

The frame pairs were aligned by laying a folding ruler against the high edge on the side bevel and lining this up with the outer edge of the bottom. One wants the frame to lie slightly inside the edge of the bottom, providing a bit of material to fine tune the plank bevel on the bottom with a plane. When both frames are set correctly, they are clamped tightly together before drilling and fastening. Also, the frames are bedded in a bead of modern caulking compound (3M 5200). Formerly the builders had used marine paint. The screws are slightly countersunk, and the heads puttied over later.

Strips

The strips lie in between each set of frames as well as between the end frames and the stem and transom. They are 1" by 2" red oak lying flat on the bottom. Their ends are trimmed parallel to the edge of the bottom leaving about a

Frames are aligned by holding a square against the outer edge and meeting the marked edge of the bottom. Note the slight bit of extra material on the bottom which will be planed flush with the frames.

1½" waterway on either side. Traditionally the strips were installed before the frames, fastened from underneath with boat nails, to hold the bottom planks together. Now Milford puts them in after the frames and screws them with stainless steel screws through the strips and into the bottom.

Stem and Transom

The stem is made from 2¼" thick red oak stock. The shape is traced from a pattern and cut on the bandsaw. This pattern has a wealth of information recorded on it, including the sheer heights for seven different dory lengths, from three to fifteen feet (the three-foot dory is called a coffee table

Milford with the stem pattern laid out on his oak stock.

(Above) A pine jig is used to scribe lines on the side and forward face of the stem marking the angle to cut for the planking. (Below) Cross-section of the stem, cut and planed to shape.

dory). The Shelburne dory has a two-piece stem, and the inner stem is mounted to the bottom. Milford had a small pattern for marking the centerline of the face of the stem as well as marking the inner rabbet on the sides. These two lines describe the bevels where the planks lie on the stem and Milford set the bandsaw to 35 degrees and cut these. The finished stem comes to a point along its forward edge.

The foot of the stem flares out providing a larger base, somewhat akin to a knee but because the stem is made of straight grained stock this foot has considerable runout of the grain. Milford noted this was a weak point when nailing and he sometimes cuts a kerf through the base of the stem and epoxies a spline to strengthen it. To prevent leaks, he also beds the stem in an epoxy/sawdust mix when fastening. The stem is aligned with the centerline on the bottom and its forward edge is brought right to the apex of the pencil lines marking the end of the bottom and fastened. We fastened the stem using stainless steel nails through the foot of the stem and into the bottom.

The transom is just over 1¼" thick, also made of red oak. Again, a pattern was used to mark its shape. The pattern also had marks for the location of the circle board, a pine piece cut to a radius that serves to strengthen the transom where holes are eventually drilled for a rope grommet. The pattern also has the 2¾" diameter hole for tracing the shape of a sculling notch. Milford bandsawed the top of the transom a full 1" longer than his mark. The finished edge would be handsawed later after planking. The sides of the transom were bandsawed at a 34-degree bevel. This was left rough and adjusted by hand as the hull was planked. The transom is supported by a knee, aligned with the centerline, bedded in epoxy/sawdust, and fastened with two screws through the

Milford aligns the stem with his marks on the bottom.

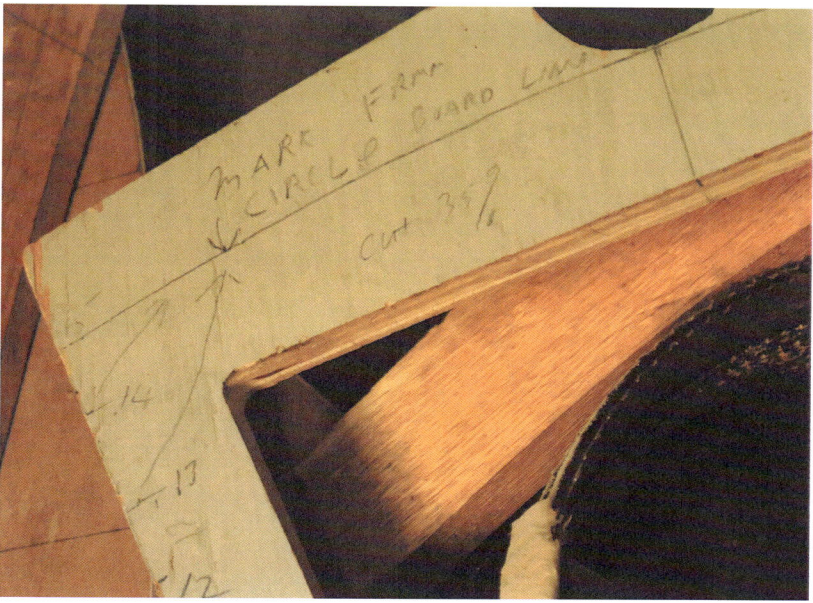

(Left) The transom pattern, Milford's plywood copy of an earlier pattern. (Above) Recorded on the transom pattern is the angle for the sides—35 degrees—and the heights for dories shorter than the 15-foot version we built.

top and two from the bottom. Like the stem, the back edge of the knee is placed so that when the transom is mounted to its back the aft face of the transom meets the very end of the marked bottom. The corners of the knee overhang the bottom but are planed flush with the planking bevel on the bottom.

The circle board overhangs the sides of the transom to ensure there is enough material when it is finally trimmed to fit. Its top edge is also beveled steeply and eventually this edge should be even with the sheer line on the pattern. Milford cut it on the bandsaw using the setting he had used to cut the transom bevels, but later said he should have cut it at 45 degrees to align closer to the sheer. It was fastened to the transom with screws through its face.

Crescent Universal Woodworker

The only stationary power tool in the dory shop is a large multi-machine with a bandsaw, joiner, shaper, and tablesaw. Milford says it was purchased in about 1920, though with its wide leather belts and clutches it is reminiscent of old water powered tools of a century earlier. It is powered by a single large electric motor. It was manufactured in Ohio and featured a 26-inch bandsaw, 12-inch joiner, 14-inch table saw and 1-inch shaper.[5] Milford's forebears in the shop said this machine's presence pushed production by five workers

With the knee and circle board fastened to the transom, Brad and Mick set the transom in place. Note the extra material left on the circle board.

---

[5] For more information visit https://t.ly/zg-hS

The only major power tool ever used in the shop is this Crescent Universal Woodworker, a multi-tool dating from the 1920s. Its adoption increased dory production to two boats per day.

Cricket checks the plank angle on the bottom using a short straightedge laid against the frame.

to over two boats per day. The only other stationary power tool we used was Milford's personal planer, which was in a mill building a few hundred yards from the dory shop.

Bottom Bevel

The first step in planking is to finalize the bevel on the edge of the bottom. At this point the backbone of the boat is complete, with the transom, frames, strips, and stem all fastened to the bottom. We carefully set this assembly up on sawhorses and shaped the bevel on the bottom one side at a time. Milford showed us how to use short straightedges, about 18" long, first laying them along the edges of the frames and against the bottom to check the bevels, which we planed to shape. Holding the straightedge flat to the frame we moved it fore and aft several inches along the bottom, checking the bevel. Milford cautioned us that we couldn't move it too far because this method is only accurate in proximity to the frame. Getting the bevel right at each frame first makes it a straightforward process to fair between those points with the plane and get the full length of the bevel right.

We then bent one of our pieces of planking stock around the entire boat, first covering the bottom bevel to check for

Planking stock bent around the boat to check the fit of the planking bevel on the frames and bottom.

(Left) Cricket moving forward along the edge of the boat wrapping candle wicking around a series of tacks set perpendicular to the face of the edge. (Above) Detail showing how the candle wicking is wrapped to cover the head of the tack.

any high spots, also checking and adjusting the bevels on the stem and transom.

With the bevel finished, Milford showed us a very interesting technique to essentially pre-caulk the seam. Starting halfway up the edge of the transom (well above the waterline) we drove ¾" cut tacks, perpendicular to the bevel face, about halfway. We continued installing tacks at about 6" intervals down the transom, then along the bottom plank bevel, halfway up the stem bevel on both sides of the boat. Then we attached candle wicking, a soft spun cotton string, to the first tack at the transom, nailing it flush, and led it down to the next tack, wrapping it once around the shank of the tack going over the top (moving the wicking around the shank clockwise). When the wicking is in the 10 o'clock position the tack is nailed, then the wicking is pulled toward the next tack, covering the head of the tack just fastened. The goal is to create a tight unbroken line of caulking material with all the tack heads covered. At the final tack on the stem the wicking is led around the front of the stem (always covering the previous tack head) to the corresponding tack on the other side, and the process continues back to the transom.

The edge of the bottom completely caulked.

The Horse. The basic building jig for all of the thousands of dories built in the shop.

Milford uses a level and straightedge to sight the transom plumb, while Brad prepares to fasten props to hold it steady.

Skillet

The fully assembled backbone of the dory is called a skillet, and the final step before starting planking is to bend it to the proper shape on the building horse. In the Shelburne boat shop, as full as it is with dozens of patterns of all shapes, darkened from years of use, the most weathered and evocative tool is the horse, which sits fixed to the center of the shop floor, like some prostrate prehistoric dragon facing the double doors where finished dories were sent to sea. Milford said the horse dated back to 1880, and thousands of dories had been built on its back. It consists of a fifteen-foot center timber 10" tall and almost 8" wide, with supporting timbers at right angles down each side. The top down the center has a long, gentle concave edge which forms the rocker (curve) when the bottom is bent down against it. The curved portion is 12'5" long and when a string is stretched tight from end-to-end the rocker—the chord of the curve—in the middle measures 3⅜". Dories of all sizes, up to twenty feet, were built on this horse.

With the skillet placed on the horse, bow toward the shop door, we set two posts on jacks fore and aft to bend the bottom. Because the skillet is longer than the horse the overhangs at either end tend to run straight, so we wedged them up carefully to follow the curve of the bottom. These wedges had to be sawn so they didn't interfere with hanging the first plank. We also wedged under the horse assembly to level the skillet side-to-side. The stem and transom must be fixed perfectly plumb. Milford said at one time there was a board mounted vertically on the shop door in line with the horse and one could straddle the skillet and line it up with the stem to set it plumb. We screwed braces from the tops of the transom and stem to the ceiling joists to hold everything absolutely steady.

Milford aligns the stem plumb by sighting the edge of the stem with a timber fastened plumb on the shop door.

With the boat held tightly to the horse with a pair of jacks, and the frames stiffened by cross spalls, a spiling batten was clamped around the hull to lay out the first plank.

Planking

Milford clamped an old batten measuring ¾" x 1½" around the frames at roughly the sheer to check that the frames all lined up. We discovered the center frame was loose on the starboard side so we used the frame pattern to get the alignment right and screwed a spreader across the frame to hold it in the proper place. All the other frames also received spreaders. The screws holding them were angled inward so they could be removed when the boat was planked up. Then with a backing dolly and hammer we tried to tighten the rivets on the midship frame.

Milford used a discarded garboard plank as a spiling batten and clamped it along the bottom on each frame. We also laid oak bars against the inside of adjacent frames so we could set a clamp between frames. Milford said his teachers always used the screw end of the clamp on the outside of the boat. "It's just how they always did it, and you didn't do it differently," he said when we asked for a reason. He checked the fit of the bottom seam again. Unlike most boats where the exact location of the planks is determined before planking—a process called lining off—in the Shelburne dory the first plank is shaped and it essentially governs the location of the remaining planks. Milford said he wanted the garboard plank to be 5" to 6" high amidships, measured on the inside (from the top face of the bottom to the top edge of the plank), and he wanted the plank to run as high as possible at the stem and transom. Since the flitch sawn material is random, he looks for his widest stock and orients the plank with the widest part of the flitch forward.

When Milford was satisfied with the way the spiling batten looked we marked the heights on the stem and transom. We then bent our planking stock around the boat and traced the shape of the bottom on it. It turned out our pattern was larger than our stock, so we reduced the height at the stem and transom and sprung our narrow batten through the two points and our midship height. With the plank shape marked, Milford cut it out using the tablesaw.[6]

---

[6] Milford cuts curved shapes on the tablesaw freehand, a technique many boatbuilders practice. Readers should know this is inherently dangerous due to the risk of kickback. A safer method would be to lay the plank on sawhorses and cut the curve with a circular saw.

A batten is clamped to planking stock and sprung through heights at the stem, midship frame, and transom. The line traced off the bottom is visible along the left edge of the plank. Once cut the plank is used to trace out an exact match.

The pine pattern used to mark the face and edge of planks. A flat bevel is planed to these marks for the dory lap.

Once cut to shape the first garboard is used as a pattern to mark the other one.

The upper edges of the garboard planks are first planed square, then Milford uses a small pine pattern to mark the plank lap: a centerline along the upper edge and a line 1¼" on the outside face parallel to the upper edge. Milford commented that the Lunenburg builders use a 1½" wide lap. With the planks clamped flat to the workbench the laps are first roughed out with a drawknife and then finished with hand planes. The gain at either end of the lap, where the bevel tapers to a feather edge, is 18" long.[7]

Milford screwed a block to the side of the stem to aid in clamping. He also ran a bead of 3M 5200 caulking at the bow and stern, about a foot along the edge of the bottom and the same distance up the stem and transom. These forward corners of the garboard seam are the most susceptible to leaks and this bit of caulking provides some insurance. He likes to clamp planks starting at the stem, bending the planks around the boat and clamping at the transom, then clamping amidships. All planks are fastened at the stem and transom with 2" galvanized boat nails. After the garboard, all subsequent planks are fastened to the frames at the plank laps only. Along the bottom we used 3" galvanized wire nails, but here only a few nails are used initially, spaced just over a foot apart. Building the boat right side up on the horse makes accurate nailing into the bottom difficult, so Milford prefers to put the remaining nails in after the boat is fully planked and rolled over. Pilot holes are drilled for all fastenings and they are also countersunk. A heavy backing iron is always held behind a timber when driving nails.

With the garboards hung on the boat, Milford then walked to the bench with a large scrap of planking and began laying out the remaining planks widths. He first drew three scales, full-size, representing plank widths at the transom, stem, and amidships. These scales laid out the distance from the *bottom* edge of the garboard lap to the sheer. It is very important to note the stem and transom scales are measured *along* those timbers (angled viewed in profile). The midships measurement is the vertical distance along the midships frame. Milford's goal is to create even, visible plank widths for the remaining planks, therefore he must take into account the laps and also the rub rail, which covers the top edge of the sheer.

When he finished deriving our remaining planks widths, Milford told us historically the plank count could vary. Our boat would have four planks, but if the planking stock were narrower, it would not be unusual to see a fifteen-footer with five planks per side. He said a 13-foot dory might have just three planks. Amazingly, he said during the days when boats were being built as quickly as possible for the fishing fleets, dories might be built with three planks on one side and four on the other.

The next plank above the garboard is called the first, or broad strake/plank. We took our next widest piece of stock,

---

[7] In lapstrake boatbuilding the gain is the end of the plank bevels, where the lap tapers to a feather edge. When two planks are joined at the gains their edges become flush.

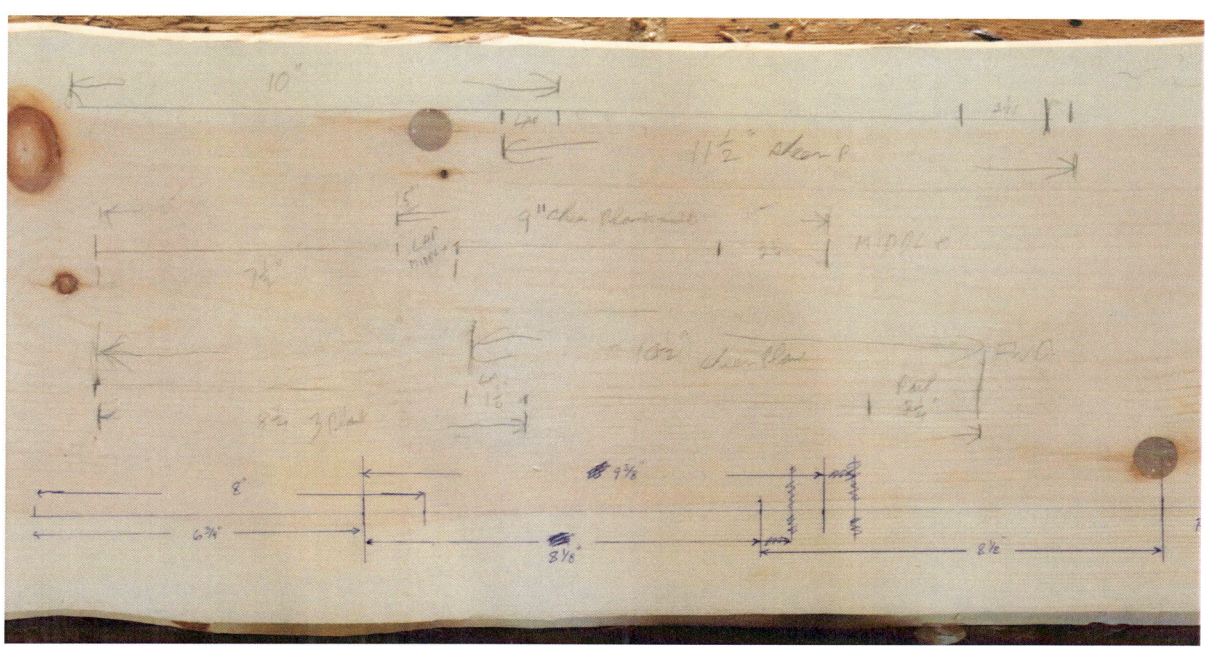

Milford's full-size layout of the plank widths at, from top to bottom, the transom, midships frame, and stem for the top two planks. The author's notes are in blue at the bottom. These are not the final dimensions but illustrate Milford's process. For each scale the upper dimensions represent actual plank widths (including material at the laps and rub rail), which the builder can transfer to the transom, midships, and stem, while the lower dimensions show how much of the finished plank will be visible. The goal is to create nearly equal values at each station for the *visible* planking.

and with the widest part of the flitch oriented aft (opposite to the garboard) we clamped it on the boat. Normally building a lapstrake hull a builder would make sure the planking stock covered the entire lap on the garboard, but Milford was only interested in covering the top edge. Boatbuilders trace the shape of the lower edge of each plank off the top edge of the preceding plank. What they must remember is to add to this line the width of the lower lap. Since Milford has already lofted his plank widths, including the upper and lower laps, his traced line becomes the bottom edge of the plank. He also traced the transom, midships frame, and stem. With the plank laid on the bench he marked the widths he had lofted

Material for the second plank, called the broad strake, is clamped onto the boat. The dory lap in the garboard is clearly visible.

from his traced edge along his transom, frame, and stem lines. He connected his three points (upper edge of plank) with a batten, drew the curve, and the plank was ready to cut to shape. Having heard his teachers' stories of how they learned, he said this method of spiling planking had been standard practice in the Shelburne shop since at least 1917.

Once shaped the gauge was used to lay out the laps, the lower lap on the inner face of the plank and the upper face on the outer. With the laps and gains cut, the starboard plank is clamped to the boat. The laps are carefully checked for gaps. If a lap is open on the outside of the hull one can remove the plank and carefully plane material off the top edge of the garboard lap on the boat. This is how we made most of our adjustments building the Shelburne dory. Once the lap fits and the plank is securely clamped Milford begins fastening, nailing first at the stem, then the transom, and finally the frames. The laps are then nailed with 1" galvanized boat nails. Again Milford uses a pattern and draws a line along the center of the lap with about a 3" gap between nails. Holes are drilled alternating just above and below the line and countersunk slightly. A backing iron is used to support the planking as the nails are driven through. The heads remain slightly proud and the tips come through the planking about ¼" and are bent over. The gains fore and aft, which are thinner, get shorter nails, ¾" long.

Traditional boat nails have a machine cut shank that tapers them, but the faces of these cuts are angled, so the cross section of the shank is a trapezoid. Milford explained we should insert nails with the narrow part of the trapezoid facing down then twist the nail slightly forward. The idea was to present two corners of the shank to the grain so the nail would cut through the wood and not act as a wedge and cause a split. For the lap nails, when bending the tips the backing iron is held against the head and the tips are bent forward and down at a 45-degree angle to the waterline.

With the starboard broad strake on we carefully transferred its height at the transom and stem and marked those heights on the port side of those timbers. We also measured its height on the midships frame and transferred it to the port side. We then clamped our planking stock for the port broad strake on the boat just as we had done for starboard and traced the bottom edge only. We shaped the lap, hung it back on the boat, checked the fit, and adjusted as needed. When the lap fit we clamped the plank in place and transferred the heights at transom, midship frame, and stem to the plank. The plank came off the boat for the final time, the three points connected with a batten, drawn and cut, and the upper lap shaped. Then it was ready to be fastened on the hull.

The remaining two strakes were marked, cut, and finished the same way, always starting with the starboard plank, which was cut using Milford's lofted dimensions, followed by the port plank, which was marked as described above to match the height of the starboard plank. We also learned that the tops of the frames should be sighted carefully at this point. If needed, they can be straightened and clamped to subsequent planks before fastening.

Staves

With three planks hung on the boat Milford made what he called staves to serve as additional frames in the gap between the forward and after frames and the ends of the boat. He used ⅝" plank scraps and used a pattern to trace their shape. The pattern is 33" long and 2⅜" wide at the top and 1¼" wide at the bottom. Milford cut them on the bandsaw. He placed them by eye, wide end to the bottom, setting their bottoms just aft of the stem and forward of the transom knee. He said he didn't square them off the bottom but instead steps back from the boat and sights them roughly parallel to the adjacent frame. Checking with a square later they are quite close to right angles from the bottom. Milford nails from the outside with a 2" cut nail set 3" above the bottom and at each lap. The points get clenched on the inside.

Sheer Planks

Before the sheer planks can be spiled the ends of the circle board have to be trimmed. Milford ran a batten around the frames across the top of the circle board and traced the angle, then he passed the batten underneath the board and scribed

The types of galvanized nails used to fasten the Shelburne dory, from top to bottom: nail for fastening the garboard to the bottom; fastening planks to stem, transom, and frames; fastening through the lap into the stem and transom; clench nailing the plank laps; and clench nailing the plank laps at the gains.

Fastening the staves in the stern. Another pair was installed between the forward frames and the bow.

it again. Next, he hand sawed the board ends just wide of the lines so they could be planed to fit.

Now the sheer planks can be traced and spiled like the preceding ones, but with one difference. Milford adds ¾" of material above his three upper marks. He said he likes to have this extra width aft as he adjusts the sheer to meet the top of the circle board. Forward he said he might make a judgement and raise the sheer a bit at the stem for looks. When cut to shape the sheer planks are the straightest of any planks on the boat.

Scribing the ends of the circle board top and bottom in order to trim to length. The sheer planks will lay over the ends.

Gunwales

For a boat this size the gunwales are stout: 1½" x 1½" red oak. We walked down the waterfront to a historic mill building where Milford had lumber stored, as well as his planer. He selected a rough sawn plank 1¾" thick and he nailed a 2" wide batten along it in a very slight curve. With his circular saw set to a 15 degree bevel he cut out a pair of gunwales. We then made multiple passes through the planer taking material from all four sides until the gunwales were 1 ½" parallelograms. We shouldered the material and headed back to the dory shop, where the two gunwales were tied together and dropped in the harbor. We soaked them for the remainder of the day, about three hours, then stored them in the shop for more soaking the next morning. Milford wanted them immersed for at least eight hours total. "Builders always milled out their gunwales for each boat, never ahead of time," he explained, "Otherwise they dry out and twist."

Cutting the frame tops. The gunwale will rest on top of the frames.

The gunwale buster, an oak wrench with a cutout that fits over the gunwale and allows a helper to twist the gunwale as it's clamped inside the hull.

Using a small offcut of the gunwale Milford laid it against the inside face of the planking at the sheer and traced a line on both sides of each frame and staves. Then he took a short saw and began trimming off the tops of these timbers where the gunwale will lay. The short saw is necessary since this cut is inside the curve of the planking. He showed us a very old, short saw he said was used in the shop for just this purpose. He also told us the grain direction of the gunwale did not matter.

Bending the gunwale in place requires at least two people. Milford has a thick piece of wood he calls a "gunwale buster," like a wrench, which one person uses aft to twist the timber while another person bends it inside the hull, starting at the bow. First he got the gunwale clamped forward and while the end was held at a twist aft Milford scribed the angle where it would fit against the stem. The gunwale was then removed from the boat and the angle sawn to shape. The process was repeated to check the fit and he re-scribed and re-cut the angle. The last step is tapering the gunwale forward to about half its width so the end lies against the side of the stem. At this point it is clamped inside the sheer plank back to the midships frame. But aft it rises out of the boat, where someone continues to twist and hold it with the gunwale buster. Milford made four or five passes with a handsaw between the side of the stem and end of the gunwale to get the fit perfect. One must be very careful tapping the gunwale forward at this point not to start to drive the sheer plank away from the stem. Once it fits Milford sets a 2" nail carefully through the sheer plank, the forward bevel of the gunwale, and into the stem.

The aft end of the gunwale is then pulled inside the boat and a first rough cut is made to length. The end of the gunwale has a steep bevel where it lies against the face of the circle board. Milford took his time with this as he didn't want to cut it too short. Once cut the gunwale is bent inside the planking and Milford began making a series of passes running the handsaw through the joint at the circle board, each pass getting the fit of the gunwale a little better. When finished a 2" stainless steel screw is driven through the top of the gunwale and into the circle board along with screws through the sheer plank into the gunwale up to the aft frame. Later these would be replaced by nails.

Next, Milford sighted carefully along the gunwale looking at the fairness of its curve. Where he saw high spots he ran his saw through the joint between the gunwale and frame heads, adjusting the gunwale until it was to his liking. The gunwale gets nailed off through the sheer planks but Milford first marked off points 15" and 20" aft of the

The gunwale is fit first at the stem, then twisted and brought carefully inside the sheer plank.

Making one of the final cuts fitting the gunwale to the face of the circle board. This takes several passes, each making the fit progressively better.

forward, forward quarter, and aft quarter frames and said we should avoid nailing in these areas because it was the location of the thole pins and we would be drilling holes there.

We fastened the starboard gunwale at the end of our seventh day in the shop. We set a new brace lower on the midships frame and removed all the other braces. When we came in the next morning Milford said he regretted doing this, because the hull did close slightly with the braces gone. He said he normally leaves all the braces on until both gunwales are installed. He thought about trying to force the hull out slightly but decided against it, so we went ahead and fitted the port gunwale. Again, lots of saw fitting to get it right. With both gunwales in and faired we used drawknives and planes to bring the sheer plank down flush with them.

The final fastening for the gunwales are galvanized sheet metal clips that attach on the inside face of the frame tops, run over the inside face of the gunwales, and are bent over and mortised into the top. First Milford used a chisel to

Metal clips are mortised flush to the top of the gunwale, nailed to the frames, and trapped by the cap rails.

bring the inside edge of the frames flush with the gunwale, chiseling a bevel at the top inch or so. The top face of the gunwale gets mortised just enough to receive the clip. The tops will not be nailed but trapped by the cap rails, which are installed next. The clips are slipped in and nailed only after the cap rails are installed, otherwise it would be impossible to plane the cap rails' inside face.

Red oak cap rail stock is laid on the gunwale, traced to shape, then cut with a bevel on the bandsaw.

Cap Rails

The rough materials for the cap rails are three-to-six foot long planks of red oak ⅝" thick and about 11" wide. Starting with a six-foot plank Milford scribed the forward edge to fit against the side of the stem and then traced their shape off the gunwale and sheer plank. Aft he cut the top of the transom in line with the sides and tops of the gunwales because the cap rail will run out and be cut off flush with the back of the transom. He bandsawed both edges at 15 degrees, cutting about ⅛" wide of the line. The excess would be planed flush after the rails are fastened. We used 1¾" galvanized wire nails to fasten it. The first nail is angled, passing through the cap rail and into the side of the stem. Similarly the last nail at the stern is angled into the transom. The remaining nails are in staggered pairs about 8" apart, and they enter only the gunwale, not the sheer plank. When nailing we held the backing iron under the gunwale to support it.

Milford cut the joint for the forward section of cap rail at a spot forward of the first thole pins. The next cap rail section is the shortest: about five-feet long, because the greatest curve in the sheer is amidships. To make the joint he lay the next piece of cap rail stock on top of the first and sawed through both with the saw tipped aft about ten degrees, with the saw cut aligned roughly athwartships. He then puts the joint together and runs a finer saw through it to get his final fit. For the third and final piece of cap rail aft, Milford tips the saw in the opposite direction when cutting the joint. All pieces are nailed on with the same pattern being sure to stay clear of where the holes for the thole pins will be drilled.

Once the cap rails' edges are planed flush the clips can be fit in their mortises and fastened. Each clip gets two 1¼" galvanized wire nails: one in the frame end and one in the inside face of the gunwale.

Bow Support

Two triangles of 1" pine were screwed to the inside corners where the stem and gunwale intersect. Their purpose is to strengthen this part of the hull where the bowline is attached through a pair of holes drilled through the supports and planking. Milford used a pattern that gave him a rough shape for the supports, then he scribed them for a tight fit. They run about a foot down either side of the stem and both visible edges get a ¼" chamfer. Each is fastened with four 1¼" #10 stainless flat head wood screws, being careful to avoid putting a fastening where the holes will be drilled.

Chock

The chock is a small piece of pine that is fastened to the inside face of the stem and lies between the bow supports. Milford milled it from 1" pine, first cutting the sides at a 45 degree angle on the bandsaw and doing the final fitting with a hand plane. "I normally screw up at least one making these," he said, as he took his first look at where he needed to plane the bevels. He starts with a piece about 16" long. The finished chock will run down the back of the stem from the cap rail about 9½". The bottom end is rounded and nailed to the stem with a pair of 2" boat nails. Once fit the excess is sawn off.

Milford used a 1" spade bit working from inside the hull to drill the holes for the bowline. The edge of the holes just touches the outer edges of the chock about 5" below the cap rail and passes through the bow supports. He drills until the spur just appears on the outside then finishes drilling from the outside in. He has a large countersink to relieve the corners of the holes. The two holes in the stern are drilled square to the face of the circle board, centered 1¾" in from the top corners and squared down 1¾". Traditionally dories would have rope grommets fashioned through these holes fore and aft so the boats could be swung off fishing schooners with davits.

Trimming the Transom

Milford laid his handsaw flat on the cap rails and let this angle guide him as he sawed off the top of the transom. The resulting shape is a very shallow peak. Earlier we had cut a U-shaped notch on the centerline of the transom for a sculling oar and after the transom was trimmed this sculling notch was roughly a semicircle.

Caulking and Fastening the Garboard

At this point the hull was turned over and set on low horses. We planed the bottom edge of the garboard almost flush with

Bow supports on either side of the stem, inside, and the chock provide extra strength at the bow, where a bowline is run through the two holes.

the bottom. Meanwhile Milford drilled for the remaining fastenings for the garboard. Originally we had tacked the garboards on with a minimum of fasteners, one every 14" to 16". Milford drilled holes so the final nail spacing is 3" to 4". Before driving these nails, however, Milford had to caulk the garboard seam.

Milford has a remarkable caulking wheel with a long handle he can use two-handed. He had told us earlier when early dories were built without glued bottoms a strand of caulking could be forced into a tight seam without any caulking bevel. Caulking wheels are normally designed to be used one-handed, but the advantage of the longer handle

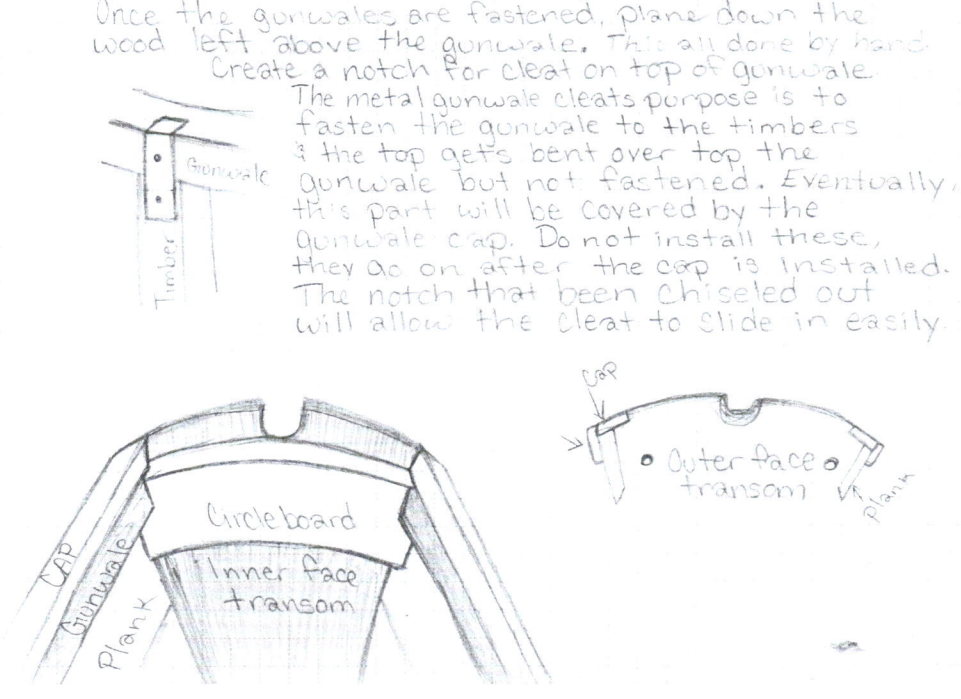

Sketch showing the transom finished, detailing the circle board, gunwale and cap, from Cricket Rust's notebook.

Using a unique, two-handed caulking wheel to force caulking into the garboard seam.

was immediately obvious. Milford could easily exert tremendous pressure driving caulking into the seam. In a few spots he ran a second strand.

We then drove the nails but did not set them. Milford insisted on setting them himself, working in a specific sequence. Starting at the stern and moving forward he first lightly set every other nail, then working back to the stern he set the remaining nails. Then he worked toward the bow again and repeated the sequence, setting every other nail and so on. He finished with one last pass setting each nail. He explained this process pulls the plank in slowly. Mick added that nailing a closely spaced row of fastenings like this increases the chance of splitting the plank, and taking the process slowly is a form of insurance.

Using a string we measured the rocker of the bottom: 4¾".

A bevel was planed at the heel of the transom to relieve the sharp corner there. This is a holdover from the era when dories were carried on fishing schooners. A sharp edge here would inevitably be damaged when a boat was launched and hauled.

Stem Band

At the bow the rough-cut ends of the planks were planed to create a smooth, flat face for the stem band. Milford sawed this from a ⅝" plank of red oak, beveling the sides. He set the bandsaw's bevel by eye, but his cut measured 22½ degrees. We bent the stem band to fit several times, looking for high spots in the face we had planed. Once it fit without gaps we drilled for the fastenings, ten 1½" #8 stainless screws and fastened it in a bedding of thickened epoxy.

Rub Strips

Milford milled two 2½" wide pine strips about fifteen feet long. One was beveled 50 degrees at the forward end to fit under the base of the stem band, running down the center of the bottom, while the other was cut in half, given 45-degree bevels on the ends, and placed outboard either side of the

Pine rubbing strips protect the bottom planks from grounding.

Cutting the nose. The two triangular gaps were filled with an epoxy/sawdust putty and the entire face sanded smooth.

center strip. All three strips were screwed to the bottom using 1¼" #8 screws in a staggered pattern about every 14", with a group of four screws fastening all ends.

Before turning the boat upright, we filled all fastening holes, along with the caulked garboard seam, with oil-based glazing compound. Once the compound is dry the entire hull is sanded.

Rub Rails or Ribbands

According to Milford rub rails were traditionally oak to withstand abrasion, but because it became too difficult to find material long enough, his teachers switched to pine. Piecing the rub rails together was problematic as joints would tend to snag things in use. Our rub rails were sawn from ⅝" x 2⅜" pine, milled straight, and were easily edge set to follow the curve of the sheer. Milford said they used to fasten them with boat nails but he uses screws now because it pulls them tight to the hull.

Nose

At the top of the stem the bow is cut off at a downward angle, something Milford called the nose. First, he laid his saw flat along the cap rail and, just like he did aft, he sawed along that rising angle through the stem head. After cutting both sides the top of the stem had a very shallow peak. He then cut the tip of the stem off angling his saw down about 30 degrees. The cut left two small cavities formed by gaps between the ends of the gunwales and the stem. We filled these with an epoxy/sawdust mixture and when dry sanded the entire nose smooth.

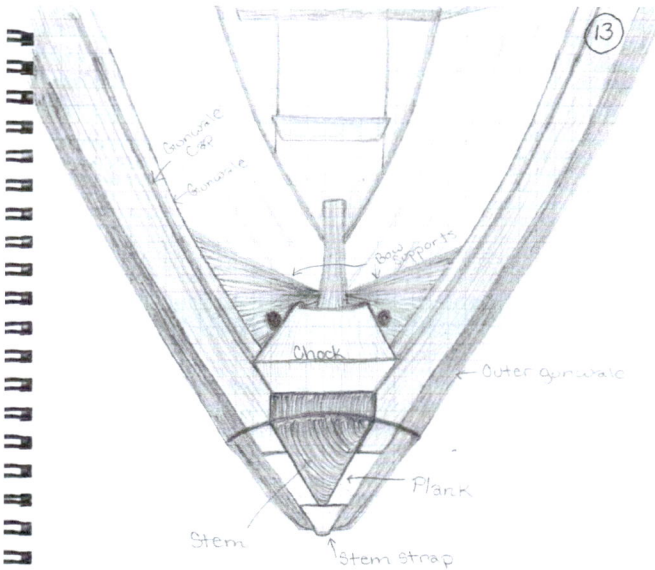

Sketch of the final finished bow, from Cricket Rust's notebook.

Thole Pins

Milford had a supply of thole pins. They are hardwood and just under ¾" in diameter. He drilled ¾" holes and said the pins should fit loosely. Three sets of holes are drilled in each gunwale for thole pins, all measurements from the centerline of the respective frames: 15½" and 20½" aft of the forward frames; 15" and 20" aft of the forward quarter frames; and 15" and 20" aft of the aft quarter frames. The hole should be drilled through the center of the gunwales matching the angle of the inner face (15 degrees). The layout is tricky because the tops of the gunwales are now covered by the cap rail. Milford measures ¹³⁄₁₆" in from the inside corner of the cap rail and punches a hole with a nail to set the spur of

The riser pattern hooks over the cap rail and a mark is made on each frame for the top of the thwart risers. The distance is 10" and the pattern is labeled saying it is for both 14-foot and 15-foot dories.

Fastening the thwart riser.

The thwarts are notched to fit over the frames. The centerline of the thwart aligns with the centerline of the frames, therefore the notches are offset to either side of the centerline.

his spade bit in. Mick kneeled on the floor holding his ruler against the inside of the gunwale to sight Milford's drill.

Risers

The thwart risers are cut from ⅝" pine and laid out with a pattern. The top edge has a noticeable curve, cut on the bandsaw to a 20-degree bevel so the thwarts rest flat against it. It is 10' 6½" long, 2¼" wide at the ends and 3" wide in the middle. The bottom edge has a very slight curve and is cut square edged. Milford also has a pattern that hooks over the cap rail and lies along the inner edge of the frames. With it he marks the location for the tops of the risers. The last brace is finally removed from the center frames. The risers land on all the frames, centered on them fore and aft, and Milford starts by setting one screw (1¼" #8) amidships, then to the forward quarter frame, edge-setting the riser to the line and setting a screw, then to the aft quarter frame, and so on until the risers are screwed at every frame. He then adds a second screw at the quarter frames.

Thwarts

Poplar used to be the wood of choice for thwarts in the dory shop, but for our boat Milford went to the local lumberyard and bought some spruce planks, which he milled to 1¼" x 8½". Three thwarts are centered on the centerline of the forward, forward quarter and aft quarter frames. Milford cuts each thwart longer than needed and draws a centerline on the top face. Milford likes to fit the thwarts like how he fit the gunwale: cut the piece long and then trim several times until it fits. First, he drops the thwart in place aligning the centerline of the frames with the line on the thwart. On the port side of the boat the line will meet the forward face of the frame; on the starboard side it aligns with the after face. He then sketches out where he will notch the thwart for the frame. He cuts these notches on the tablesaw by pushing the thwart into the blade to his sketched line and withdrawing it. At the bench he finishes his notches with a chisel. The ends of the thwarts are beveled parallel to the planking but Milford said they should not touch the planks.

When finally fit the thwarts rest on the risers and the notches have a loose fit. They are not fastened, harkening back to the days when on board fishing schooners thwarts would be pulled from the dories so the boats could be compactly stacked one inside another. This is one reason why the gunwales on dories seem so oversized; these boats lack any athwartship framing to support their shape.

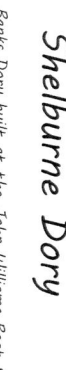

Brad, Cricket, and Diane take the completed Shelburne Dory out for a spin.

Conclusion

Graham McKay

The dory building tradition practiced at the Shelburne Dory Shop is significant for many reasons. Primarily, because it is the least watered down, most authentic dory building tradition known to exist. That there are so few people between the shop's founders and Milford would indicate a high degree of originality in the tradition. With every generation there are updates and improvements in the process as skill and technology dictate, and this is the case as well with Milford, although he has tried to maintain the process just as the old feller did. In the text we pointed out modern materials Milford has adopted, such as epoxy and stainless-steel screws, but in each case Milford was able to tell us the precise methods his predecessors used.

One of the other miracles that has kept this tradition so pure is the extraordinarily long life and tenure of Sidney Mahaney, who worked at the shop from 1914 to 1990 (76 years) and lived to be over 100 years old. Another miracle is that there are only three degrees of separation from Milford to John Williams, who opened the shop in 1880. Williams ran the shop until his passing in 1939 when Charles Wyman took over for twenty years. until Bill Cox, who overlapped with Milford, purchased the shop in 1959 and ran it until its eventual closing in 1971, employing Mahaney the entire time. When the shop reopened as a museum in 1983 Mahaney came back to work until his son Curtis took over as the builder. Milford learned from Curtis and had old Bill Cox still coming by to keep him honest. Undoubtedly Mahaney altered the process some during his lifetime, but it's unlikely that any individual drastically changed the process.

The other prominent traditions, such as that practiced at the Lunenburg Dory Shop, and at Lowell's Boat Shop in Massachusetts, have chains of knowledge that either have broken or missing links. At Lowells for instance, the design of the dory changed around the time of World War II when dories ceased to be used for trawl fishing from schooners, and just became an inexpensive utility boat. It is also a time when the shop slowed to a very minimal production with a skeleton crew because the boat builders were conscripted to work at the shipyard in Portsmouth, New Hampshire for war effort. No doubt this altered things from pre-war to postwar time. This is one instance of many in the history of Lowell's Boat Shop (Hiram Lowell & Sons) that has led to the dilution of the tradition. The history of the Lunenburg shop, which is the youngest of the three mentioned having been established just after WWI, is little different. It has had its own eras of going fallow, changing ownership, and not producing many, if any, dories. Each hiccup in the chain caused new builders to have to learn from old if they were still around or perhaps from just looking at old boats, which, unfortunately, don't last very long. To that end, it is very difficult to find an old dory built by any shop...and by old I mean older than perhaps 50 years and certainly not anything built before 1900; those dories are most certainly extinct.

The small number of builders over time at the Shelburne Dory Shop is one way the tradition has remained so true to the original. One other reason is that the shop itself has not been improved much, if at all, over time. The machinery, the tools, even the power and modern conveniences have not infiltrated the shop to any great extent. The few power tools that are used are ancient. There is very little light to work from other than natural light and there are a number of original patterns that are still used to produce these dories. Not unlike some of the other shops, there are patterns on the walls and stored in the overhead whose use is unknown to Milford, but they remain in their place because they have always been there. Patterns can be confusing as they can often be created for a one-time use. In the case of the patterns that are regularly used by Milford at the Dory Shop many of the marks and writing on them are illegible. Also, like any tradition, there is so much of the process that was handed down verbally from generation to generation, builder to builder, and even those with the best of memories can't always recall 40- or 50-year-old details with a high degree of exactitude. Milford's heavy reliance on patterns, for even the smallest details of construction, do help to primarily freeze his techniques, linking them precisely to past boatbuilders. Milford told us he had copied the oldest patterns as their notations became illegible. These patterns are undoubtedly the oldest artifacts in the shop, used repeatedly again and again, they are the surest evidence of the consistency of the process we recorded.

There are some parts of the tradition that have been lost, because they are no longer in use. Because the Shelburne Dory Shop has not had to produce a dory to go on the deck of a schooner for fishing in at least half a century, if not more, some of these features are no longer added to the boat, and therefore, that part of the tradition falls away. One such detail is the reinforcing of the rail forward where the Hurdy Gurdy would sit. Other items are pen boards, and other sundry fittings related to commercial fishing. Interestingly enough, given the age and originality of the shop itself, there are very few signs of the structure being used as a means of building dories. Other than the horse, which has seen thousands of boats produced on it, the benches and such don't appear to be particularly modified in any way for the production of dories. The rudimentary system for creating the miter for the frames has the appearance of being engineered and knocked together hastily by a 19th century builder and never improved upon since. No doubt during the height of production things looked much different in the shop, and the various workers kept to their stations and those stations were likely built to suit. In the more modern age (i.e. the second half of the 20th century) and certainly much more since the dory shop became part of the museum in Shelburne, the forward end of the upstairs shop has been dedicated as the building space with much of the other areas of the structure being filled with old dories and exhibit space. Understandably, some of the traditions had to have been lost when dory production ceased on a massive scale and reverted to a single builder building a single dory.

Even with the nearly pure chain of knowledge that exists in Shelburne through Milford we can lamentably conclude that much of the process, as it was carried out in the years of production, have been lost to time. The actual construction of the dory may be largely unchanged—and recorded here—but the more abstract parts of production like the procuring and storing of materials, purpose made tools and shop spaces, jigs and pattern uses, etc. are no longer known. Despite this, there is still so much of the old history and techniques still practiced, and no doubt more could be gleaned from Milford than is recorded here. However, it would take years of working alongside him to really learn the nuances. Thanks to Milford's generosity, we believe we have recorded the process to as detailed a degree as possible given the amount of time we worked with him, and we hope to do the same with the other dory traditions. The hope is that this boat and the time-honored methods used to build it will be perpetuated through this publication and that others will be inspired to build these dories "just as the old feller done it."

Bibliography

John Gardner, The Dory Book, International Marine Publishing, 1978.

Otto Kelland, Dories and Dorymen, Robinson-Blackmore, 1984.

Lewis Robertson, Shelburne and the Gloustermen, Shelburne Historical Society, 1977.

Howard Chapelle, American Small Sailing Craft, W.W. Norton, 1951.

Wooden Boat Magazine, Brooklin, Maine, various issues, 1974–present.

Dory Shops

A short and incomplete list of dory shops still producing the occasional traditional dory.

Amesbury, Mass. Lowell's Boat Shop. Graham McKay

Gloucester, Mass. Dory Shop, Geno Mondello

Shelburne, Nova Scotia, Dory Shop Museum, Milford Buchanan, Anne Poirier

Lunenburg, Nova Scotia, Dory Shop

The Big Boat Shed, Andrew Rhodenizer, David Westergard

Winterton, Newfoundland, Wooden Boat Museum of Newfoundland and Labrador, Jerome Canning

St. Pierre, St. Pierre And Miquelon, Les Zigotos, Gerard Helene

www.ingramcontent.com/pod-product-compliance
Lightning Source LLC
Chambersburg PA
CBRC100222100526
44590CB00008B/146